YOUR KNOWLEDGE HAS VALUE

- We will publish your bachelor's and
  master's thesis, essays and papers

- Your own eBook and book -
  sold worldwide in all relevant shops

- Earn money with each sale

Upload your text at www.GRIN.com
and publish for free

C. P. Kumar

Impact of Climate Change on Groundwater Resources

GRIN Verlag

Bibliografische Information der Deutschen Nationalbibliothek:

Die Deutsche Bibliothek verzeichnet diese Publikation in der Deutschen National-
bibliografie; detaillierte bibliografische Daten sind im Internet über http://dnb.d-
nb.de/ abrufbar.

Imprint:

Copyright © 2014 GRIN Verlag GmbH
Druck und Bindung: Books on Demand GmbH, Norderstedt Germany
ISBN: 978-3-656-75552-1

This book at GRIN:

http://www.grin.com/en/e-book/281606/impact-of-climate-change-on-groundwater-
resources

IMPACT OF CLIMATE CHANGE ON GROUNDWATER RESOURCES

C. P. Kumar

Scientist 'F', National Institute of Hydrology, Roorkee – 247667 (Uttarakhand)

Abstract

Climate change poses uncertainties to the supply and management of water resources. The Intergovernmental Panel on Climate Change (IPCC) estimates that the global mean surface temperature has increased 0.6 ± 0.2 °C since 1861, and predicts an increase of 2 to 4 °C over the next 100 years. Temperature increases also affect the hydrologic cycle by directly increasing evaporation of available surface water and vegetation transpiration. Consequently, these changes can influence precipitation amounts, timings and intensity rates, and indirectly impact the flux and storage of water in surface and subsurface reservoirs (i.e., lakes, soil moisture, groundwater). In addition, there may be other associated impacts, such as sea water intrusion, water quality deterioration, potable water shortage, etc.

While climate change affects surface water resources directly through changes in the major long-term climate variables such as air temperature, precipitation, and evapotranspiration, the relationship between the changing climate variables and groundwater is more complicated and poorly understood. The greater variability in rainfall could mean more frequent and prolonged periods of high or low groundwater levels, and saline intrusion in coastal aquifers due to sea level rise and resource reduction. Groundwater resources are related to climate change through the direct interaction with surface water resources, such as lakes and rivers, and indirectly through the recharge process. The direct effect of climate change on groundwater resources depends upon the change in the volume and distribution of groundwater recharge. Therefore, quantifying the impact of climate change on groundwater resources requires not only reliable forecasting of changes in the major climatic variables, but also accurate estimation of groundwater recharge.

A number of Global Climate Models (GCM) are available for understanding climate and projecting climate change. There is a need to downscale GCM on a basin scale and couple them with relevant hydrological models considering all components of the hydrological cycle. Output of these coupled models such as quantification of the groundwater recharge will help in taking appropriate adaptation strategies due to the impact of climate change. This article presents the likely impact of climate change on groundwater resources, climate change scenario for groundwater in India, status of research studies carried out at national and international level, and methodology to assess the impact of climate change on groundwater resources.

Keywords: Climate change; Hydrological cycle; Groundwater recharge; seawater intrusion; Numerical modeling; MODFLOW; UnSat Suite; WetSpass.

Introduction

Water is indispensable for life, but its availability at a sustainable quality and quantity is threatened by many factors, of which climate plays a leading role. The Intergovernmental Panel on Climate Change (IPCC) defines climate as "the average weather in terms of the mean and its variability over a certain time-span and a certain area" and a statistically significant variation of the mean state of the climate or of its variability lasting for decades or longer, is referred to as climate change.

Evidence is mounting that we are in a period of climate change brought about by increasing atmospheric concentrations of greenhouse gases. Atmospheric carbon dioxide levels have continually increased since the 1950s. The continuation of this phenomenon may significantly alter global and local climate characteristics, including temperature and precipitation. Climate change can have profound effects on the hydrologic cycle through precipitation, evapotranspiration, and soil moisture with increasing temperatures. The hydrologic cycle will be intensified with more evaporation and more precipitation. However, the extra precipitation will be unequally distributed around the globe. Some parts of the world may see significant reductions in precipitation or major alterations in the timing of wet and dry seasons. Information on the local or regional impacts of climate change on hydrological processes and water resources is becoming more important. The effects of global warming and climatic change require multi-disciplinary research, especially when considering hydrology and global water resources.

The Intergovernmental Panel on Climate Change (IPCC) estimates that the global mean surface temperature has increased 0.6 ± 0.2 °C since 1861, and predicts an increase of 2 to 4 °C over the next 100 years. Global sea levels have risen between 10 and 25 cm since the late 19th century. As a direct consequence of warmer temperatures, the hydrologic cycle will undergo significant impact with accompanying changes in the rates of precipitation and evaporation. Predictions include higher incidences of severe weather events, a higher likelihood of flooding, and more droughts. The impact would be particularly severe in the tropical areas, which mainly consist of developing countries, including India.

Coupled atmosphere-ocean global climate models (GCMs) are used to estimate changes in climate. These physically-based numerical models simulate synoptic-scale climate and hydrological processes, and are forced with greenhouse gas and aerosol emission scenarios. A wide diversity of GCMs developed by leading climate centres are available for other researchers to evaluate potential impacts of climate change. To ensure that the predictive elements from a GCM are realistic, a statistical downscaling technique should be employed to bridge the local- and synoptic-scale processes. Statistical downscaling uses a correlation between predictands (site measured variables, such as precipitation) and predictors (region-scale variables, such as GCM variables).

Changes in regional temperature and precipitation have important implications for all aspects of the hydrologic cycle. Variations in these parameters determine the amount of water that reaches the surface, evaporates or transpires back to the atmosphere, becomes stored as snow or ice, infiltrates into the groundwater system, runs off the land, and ultimately becomes base flow to streams and rivers.

Hydrological impact assessments of watersheds (and aquifers) require information on changes in evapotranspiration because it is a key component of the water balance. However, climate-change scenarios tend to be expressed in terms of changes in temperature and precipitation. Consequently, the effects of global warming on potential evaporation (or more inclusively, evapotranspiration) are not simple to estimate. Many global scenarios suggest an increase in potential evaporation, but these factors may be outweighed locally or regionally by other factors reducing evaporation. Various models may be used to calculate potential evaporation using data on net radiation, temperature, humidity, and wind speed, and sometimes plant physiological properties. The estimated effect of a change in climate on potential evaporation depends on the characteristics of the site.

Many rivers and streams that are fed by glacier runoff could be significantly impacted as a result of climate change. As glacier retreat accelerates, increased summer runoff could occur. However, when the glaciers have largely melted, the late summer and fall glacial input into streams and rivers may be lost, resulting in a significant reduction in flow in some cases.

Water resource management plans increasingly need to incorporate the affects of global climate change in order to accurately predict future supplies. Numerous studies have documented the sensitivity of streamflow to climatic changes for watersheds all over the world. Most of these studies involve watershed scale hydrologic models, of which validation remains a fundamental challenge. Moreover, outputs from general circulation models (GCM) can be rather uncertain and downscaling their predictions for local hydrologic use can produce inconsistent results. Therefore, the sensitivity of streamflow to climate changes is perhaps best understood by analyzing the historical records.

Building empirical models to link climate and regional hydrological regimes has a long history. In recent years, many researchers have used empirical rainfall–runoff model to study the impacts of climatic change on hydrology. However, applications of these empirical relationships to climate or basin conditions different from those used in the original development of these functions are questionable.

Impact of Climate Change on Groundwater Resources

Although the most noticeable impacts of climate change could be fluctuations in surface water levels and quality, the greatest concern of water managers and government is the potential decrease and quality of groundwater supplies, as it is the main available potable water supply source for human consumption and irrigation of agriculture produce worldwide. Because groundwater aquifers are recharged mainly by precipitation or through interaction with surface water bodies, the direct influence of climate change on precipitation and surface water ultimately affects groundwater systems.

It is increasingly recognized that groundwater cannot be considered in isolation from the landscape above, the society with which it 'interacts', or from the regional hydrological cycle, but needs to be managed holistically. In understanding the likely consequences of possible future (climate and non-climate) changes on groundwater systems and the regional hydrological cycle, an important (but not exclusive) component to understand is the influence that these factors exert on recharge and runoff.

It is important to consider the potential impacts of climate change on groundwater systems. As part of the hydrologic cycle, it can be anticipated that groundwater systems will be affected by changes in recharge (which encompasses changes in precipitation and evapotranspiration), potentially by changes in the nature of the interactions between the groundwater and surface water systems, and changes in use related to irrigation.

(a) Soil Moisture

The amount of water stored in the soil is fundamentally important to agriculture and has an influence on the rate of actual evaporation, groundwater recharge, and generation of runoff. Soil moisture contents are directly simulated by global climate models, albeit over a very coarse spatial resolution, and outputs from these models give an indication of possible directions of change.

The local effects of climate change on soil moisture, however, will vary not only with the degree of climate change but also with soil characteristics. The water-holding capacity of soil will affect possible changes in soil moisture deficits; the lower the capacity, the greater the sensitivity to climate change. Climate change also may affect soil characteristics, perhaps through changes in waterlogging or cracking, which in turn may affect soil moisture storage properties. Infiltration capacity and water-holding capacity of many soils are influenced by the frequency and intensity of freezing.

(b) Groundwater Recharge and Resources

Groundwater is the major source of water across much of the world, particularly in rural areas in arid and semi-arid regions, but there has been very little research on the potential effects of climate change.

Aquifers generally are replenished by effective rainfall, rivers, and lakes. This water may reach the aquifer rapidly, through macro-pores or fissures, or more slowly by infiltrating through soils and permeable rocks overlying the aquifer. A change in the amount of effective rainfall will alter recharge, but so will a change in the duration of the recharge season. Increased winter rainfall, as projected under most scenarios for mid-latitudes, generally is likely to result in increased groundwater recharge. However, higher evaporation may mean that soil deficits persist for longer and commence earlier, offsetting an increase in total effective rainfall. Various types of aquifer will be recharged differently. The main types are unconfined and confined aquifers. An unconfined aquifer is recharged directly by local rainfall, rivers, and lakes, and the rate of recharge will be influenced by the permeability of overlying rocks and soils.

Macro-pore and fissure recharge is most common in porous and aggregated forest soils and less common in poorly structured soils. It also occurs where the underlying geology is highly fractured or is characterized by numerous sinkholes. Such recharge can be very important in some semi-arid areas. In principle, "rapid" recharge can occur whenever it rains, so where recharge is dominated by this process it will be affected more by changes in rainfall amount than by the seasonal cycle of soil moisture variability.

Shallow unconfined aquifers along floodplains, which are most common in semi-arid and arid environments, are recharged by seasonal streamflows and can be depleted directly by evaporation. Changes in recharge therefore will be determined by changes in the duration of flow of these streams, which may locally increase or decrease, and the permeability of the overlying beds, but increased evaporative demands would tend to lead to lower groundwater storage. The thick layer of sands substantially reduces the impact of evaporation.

It will be noted from the foregoing that unconfined aquifers are sensitive to local climate change, abstraction, and seawater intrusion. However, quantification of recharge is complicated by the characteristics of the aquifers themselves as well as overlying rocks and soils. A confined aquifer, on the other hand, is characterized by an overlying bed that is impermeable, and local rainfall does not influence the aquifer. It is normally recharged from lakes, rivers, and rainfall that may occur at distances ranging from a few kilometers to thousands of kilometers.

Aside from the influence of climate, recharge to aquifers is very much dependent on the characteristics of the aquifer media and the properties of the overlying soils. Several approaches can be used to estimate recharge based on surface water, unsaturated zone and groundwater data. Among these approaches, numerical modelling is the only tool that can predict recharge. Modelling is also extremely useful for identifying the relative importance of different controls on recharge, provided that the model realistically accounts for all the processes involved. However, the accuracy of recharge estimates depends largely on the availability of high quality hydrogeologic and climatic data. Determining the potential impact of climate change on groundwater resources, in particular, is difficult due to the complexity of the recharge process, and the variation of recharge within and between different climatic zones.

Attempts have been made to calculate the rate of recharge by using carbon-14 isotopes and other modeling techniques. This has been possible for aquifers that are recharged from short distances and after short durations. However, recharge that takes place from long distances and after decades or centuries has been problematic to calculate with accuracy, making estimation of the impacts of climate change difficult. The medium through which recharge takes place often is poorly known and very heterogeneous, again challenging recharge modeling. In general, there is a need to intensify research on modeling techniques, aquifer characteristics, recharge rates, and seawater intrusion, as well as monitoring of groundwater abstractions. This research will provide a sound basis for assessment of the impacts of climate change and sea-level rise on recharge and groundwater resources.

(c) Coastal Aquifers

When considering water resources in coastal zones, coastal aquifers are important sources of freshwater. However, salinity intrusion can be a major problem in these zones. Salinity intrusion refers to replacement of freshwater in coastal aquifers by saltwater. It leads to a reduction of available fresh groundwater resources. Changes in climatic variables can significantly alter groundwater recharge rates for major aquifer systems and thus affect the availability of fresh groundwater. Salinization of coastal aquifers is a function of the reduction of groundwater recharge and results in a reduction of fresh groundwater resources.

Sea-level rise will cause saline intrusion into coastal aquifers, with the amount of intrusion depending on local groundwater gradients. Shallow coastal aquifers are at greatest risk. Groundwater in low-lying islands therefore is very sensitive to change. A reduction in precipitation coupled with sea-level rise would not only cause a diminution of the harvestable volume of water; it also would reduce the size of the narrow freshwater lense. For many small island states, such as some Caribbean islands, seawater intrusion into freshwater aquifers has been observed as a result of overpumping of aquifers. Any sea-level rise would worsen the situation.

A link between rising sea level and changes in the water balance is suggested by a general description of the hydraulics of groundwater discharge at the coast. Fresh groundwater rides up over denser, salt water in the aquifer on its way to the sea (Figure 1), and groundwater discharge is focused into a narrow zone that overlaps with the intertidal zone. The width of the zone of groundwater discharge measured perpendicular to the coast, is directly proportional to the discharge rate. The shape of the water table and the depth to the freshwater/saline interface are controlled by the difference in density between freshwater and salt water, the rate of freshwater discharge and the hydraulic properties of the aquifer. The elevation of the water table is controlled by mean sea level through hydrostatic equilibrium at the shore.

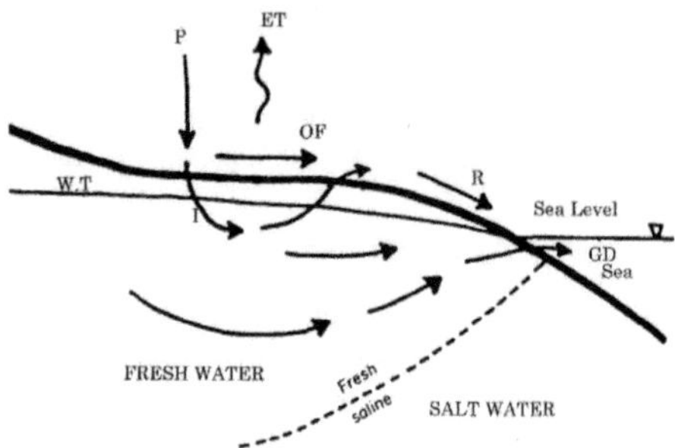

Figure 1: Conceptual Model of the Water Balance in a Coastal Watershed

To assess the impacts of potential climate change on fresh groundwater resources, we should focus on changes in groundwater recharge and sea level rise on the loss of fresh groundwater resources in water resources stressed coastal aquifers.

Climate Change Scenario for Groundwater in India

Impact of climate change on the ground water regime is expected to be severe. It is to be pointed out that groundwater is the principle source of drinking water in the rural areas. About 85% of the rural water supply in India is dependent on groundwater. India on the whole has a potential of

431 bcm/year of replenishable groundwater (CGWB, 2011), Unfortunately, due to rampant drawing of the subsurface water, the water table in many regions of the country has dropped significantly in the recent years resulting in threat to groundwater sustainability. The overexploited areas are mostly concentrated on three parts of the country. In north western part in Punjab, Haryana, Delhi, Western Uttar Pradesh where through replenishable resources is abundant but there have indiscriminate withdrawals of ground water leading to over-exploitation, in western part of the country particularly in Rajasthan where due to arid climate, ground water recharge itself is less leading to stress on the resource and in peninsular India like Karnataka and Tamil Nadu where due to poor aquifer properties, ground water availability is less.

The most optimistic assumption suggests that an average drop in groundwater level by one metre would increase India's total carbon emissions by over 1%, because the time of withdrawal of the same amount of water will increase fuel consumption. A more realistic assumption reflecting the area projected to be irrigated by groundwater, suggests that the increase in carbon emission could be 4.8% for each metre drop in groundwater levels (Mall et al., 2006). It is recommended to study the aquifer geometry, establish the saline fresh interfaces within few km of the coastal area, the effect of glaciers melting on recharge potential of aquifers in the Ganga basin and its effects on the transboundary aquifer systems, particularly in the arid and semi-arid regions.

Climate change is likely to affect ground water due to changes in precipitation and evapotranspiration. Rising sea levels may lead to increased saline intrusion into coastal and island aquifers, while increased frequency and severity of floods may affect groundwater quality in alluvial aquifers. Sea-level rise leads to intrusion of saline water into the fresh groundwater in coastal aquifers and thus adversely affects groundwater resources. For two small and flat coral islands at the coast of India, the thickness of freshwater lens was computed to decrease from 25 m to 10 m and from 36 m to 28 m, respectively, for a sea level rise of only 0.1 m (Mall et al., 2006).

Increased amount of precipitation in short heavy spells will lead to low infiltration thereby causing low moisture availability for soil. Furthermore, water management systems in the area like number of reservoirs, boreholes etc. would also modify the water availability. Global warming will also affect the water supply by changes in evaporation and ground water recharge. Finally through sea level rise, the global warming may contribute saline intrusion.

Agricultural demand, particularly for irrigation water, which is a major share of total water demand of the country, is considered more sensitive to climate change. A change in field-level climate may alter the need and timing of irrigation. Increased dryness may lead to increased demand, but demand could be reduced if soil moisture content rises at critical times of the year. It is projected that most irrigated areas in India would require more water around 2025 and global net irrigation requirements would increase relative to the situation without climate change by 3.5–5% by 2025 and 6–8% by 2075. In India, roughly 52% of irrigation consumption across the country is extracted from groundwater; therefore, it can be an alarming situation with decline in groundwater and increase in irrigation requirements due to climate change.

Warm air holds more moisture and increase evaporation of surface moisture. With more moisture in the atmosphere, rainfall and snowfall events tend to be more intense, increasing the potential for floods. However, if there is little or no moisture in the soil to evaporate, the incident solar radiation goes into raising the temperature, which could contribute to longer and more severe droughts. Therefore, change in climate will affect the soil moisture, groundwater recharge and frequency of flood or drought episodes and finally groundwater level in different areas. In a number of studies, it is projected that increasing temperature and decline in rainfall may reduce net recharge and affect groundwater levels. However, little work has been done on hydrological impacts of possible climate change for Indian regions/basins.

Existing economic growth scenarios project total power generation capacity in India to increase nine times from 96 GW to 912 GW between 1995-2100. As a result of climate change, it is estimated that approximately 1.5% more power generation capacity will be required. Increased energy demand may arise from a number of sources. For example, increases in average temperature can result in the need for space cooling for buildings, and variability in precipitation can impact irrigation needs and consequent demand for energy from groundwater pumping.

Status of Research Studies

The increase of concentration of carbon dioxide and other greenhouse gases in the atmosphere will certainly affect hydrological regimes. Global warming is thus expected to have major implications on water resources management. The observation of long-term trends in climate for many regions of the world has led to considerable research on the impact of greenhouse gases on climate. To this end, several general circulation models (GCMs) have been used to simulate the type of climate that might exist if global concentrations of carbon dioxide (greenhouse gas) were twice their pre-industrial levels. Recent GCM estimates of the projected rise in long-term global average annual surface temperature are between 1 and 4.5 $^{\circ}$C under simulated doubled concentrations of CO_2. On the subcontinent scale, there remains considerable uncertainty in the model results and it is not possible to know with confidence the fine details of how the climate will change regionally (Taylor 1997). Consequently, it is customary to use observational data as a baseline and adjust these by the GCM scenarios (Taylor 1997). Because precipitation patterns are significantly influenced by changes in the global-circulation patterns induced by climate change, regional projections for changes in precipitation under doubled CO_2 scenarios remain very uncertain.

There have been many studies relating the effect of climate changes on surface water bodies. However, very little research exists on the potential effects of climate change on groundwater, although groundwater is the major source of drinking water across much of the world and plays a vital role in maintaining the ecological value of an area. Available studies show that groundwater recharge and discharge conditions are reflection of the precipitation regime, climatic variables, landscape characteristics and human impacts such as agricultural drainage and flow regulation. Hence, predicting the behavior of recharge and discharge conditions under future climatic and other changes is of great importance for integrated water management.

Studies which consider the indirect effects derived from climate-change-induced alterations in soil, land cover, salt-water intrusion due to rising sea levels and changes in water demand are less common. These studies represent a move away from impact studies (which may be considered to be vertically integrated, in which climate change acts upon an environmental compartment) towards horizontally integrated studies in which environmental compartments interact with each other. However, they remain an incomplete assessment of the pressures facing groundwater resources associated with the direct and indirect effects of future climate and socio-economic change.

Previous studies have typically coupled climate change scenarios with hydrological models, and have generally investigated the impact of climate change on water resources in different areas. The scientific understanding of an aquifer's response to climate change has been studied in several locations within the past decade. These studies link atmospheric models to unsaturated soil models, which, in some cases, were further linked into a groundwater model. The groundwater models used were calibrated to current groundwater conditions and stressed under different predicted climate change scenarios. Some of the recent studies on impact of climate change on groundwater resources have been discussed below.

Vaccaro (1992) investigated the degree of variability in climate and its impact on future recharge predictions in a basin in the northwestern United States. In addition to historical records, climate predictions from the synthetic weather generator WGEN (Richardson and Wright, 1984) and three GCMs were considered along with two different land use conditions. The results of the study indicated that the variability in annual recharge was less under the GCM conditions than using the historic data.

Bouraoui et al. (1999) presented a general approach to evaluate the effect of potential climate changes on groundwater resources. In the current stage of knowledge, large-scale global climate models are probably the best available tools to provide estimates of the effects of raising greenhouse gases on rainfall and evaporation patterns through a continuous, three dimensional simulation of atmospheric, oceanic and cryospheric processes. However their spatial resolution (generally some thousands of square kilometers) is not compatible with that of watershed hydrologic models. The main purpose of this study is to evaluate the impact of potential climate changes upon groundwater resources. A general methodology is proposed in order to disaggregate outputs of large-scale models and thus to make information directly usable by hydrologic models. As an illustration, this method is applied to a CO_2-doubling scenario through the development of a local weather generator, although many uncertainties are not yet assessed about the results of climate models. Two important hydrological variables: rainfall and potential evapotranspiration are thus generated. They are then used by coupling with a physically based hydrological model to estimate the effects of climate changes on groundwater recharge and soil moisture in the root zone.

Rosenberg et al. (1999) studied the impact of climate change on the water yield and groundwater recharge of the Ogallala aquifer in the central United States. Three different GCMs were used to predict changes in the future climate due to anticipated changes in temperature and CO_2 concentrations. The study found that recharge was reduced under all scenarios, ranging up to 7%, depending on the simulation conditions.

Sherif and Singh (1999) investigated the possible effect of climate change on sea water intrusion in coastal aquifers. There is increasing debate these days on climate change and its possible consequences. Much of the debate has focused in the context of surface water systems.

In many arid areas of the world, rainfall is scarce and so is surface runoff. These areas rely heavily on groundwater. The consequences of climate change on groundwater are long term and can be far reaching. One of the more apparent consequences is the increased migration of salt water inland in coastal aquifers. Using two coastal aquifers, one in Egypt and the other in India, this study investigated the effect of likely climate change on sea water intrusion. Under conditions of climate change, the sea water levels will rise for several reasons, including variations in atmospheric pressures, expansion of warmer occasions and seas and melting of ice sheets and glaciers. The rise in sea water levels will impose additional saline water heads at the sea side and therefore more sea water intrusion is anticipated. Three realistic scenarios mimicking climate change were considered. Under these scenarios, the Nile Delta aquifer is found to be more vulnerable to climate change and sea level rise. A 50 cm rise in the Mediterranean sea level will cause additional intrusion of 9.0 km in the Nile Delta aquifer. The same rise in water level in the Bay of Bengal will cause an additional intrusion of 0.4 km. Additional pumping will cause serious environmental effects in the case of the Nile Delta aquifer.

Ghosh Bobba (2002) analysed the effects of human activities and sea-level changes on the spatial and temporal behaviour of the coupled mechanism of salt-water and freshwater flow through the Godavari Delta of India. The density driven salt-water intrusion process was simulated with the use of a SUTRA (Saturated-Unsaturated TRAnsport) model. Physical parameters, initial heads, and boundary conditions of the delta were defined on the basis of available field data, and an areal, steady-state groundwater model was constructed to calibrate the observed head values corresponding to the initial development phase of the aquifer. Initial and boundary conditions determined from the areal calibration were used to evaluate steady-state, hydraulic heads. Consequently, the initial position of the hydraulic head distribution was calibrated under steady-state conditions. The changes of initial hydraulic distribution, under discharge and recharge conditions, were calculated, and the present-day position of the interface was predicted. The present-day distribution of hydraulic head was estimated via a 20-year simulation. The results indicate that a considerable advance in seawater intrusion can be expected in the coastal aquifer if current rates of groundwater exploitation continue and an important part of the freshwater from the river is channelled from the reservoir for irrigation, industrial and domestic purposes.

Groundwater investigations are presently very active in Andhra Pradesh because of the urgent need for more water to meet the demands for the agricultural, industrial, and domestic purposes of the growing population in the coastal areas. This research has provided numerical simulation of the influence of surface flows, coming from the water management of the projected reservoir, on the regional groundwater behaviour in the delta aquifer. A two-dimensional finite element model, considering open boundary conditions for coasts and a sharp interface between freshwater and salt water, was applied to the aquifer under steady-state conditions for freshwater surplus and deficits at the coastline. When recharges of salt water occur at the coastline, essentially of freshwater deficits, a hypothesis of mixing for the freshwater-salt water transition zone allows the model to calculate the resulting seawater intrusion in the aquifer. Hence, an adequate treatment and interpretation of the hydrogeological data, which are available for the coastal aquifer, were of main concern in satisfactorily applying the proposed numerical model. The results of the steady-state simulations showed reasonable calculations of the water table levels and the freshwater and salt-water thicknesses, as well as the extent of the interface and seawater intrusion into the aquifer for the total discharges or recharges in the delta and along the coastline. As a result of the present hydrogeological conditions, a considerable advance in seawater intrusion would be expected in the coastal aquifer if current rates of groundwater exploitation continue and an important part of the freshwater from the river is channeled from the reservoir for irrigation, industrial and domestic purposes.

Kirshen (2002) used the groundwater model MODFLOW to study the impact of global warming on a highly permeable aquifer in the northeastern United States. Groundwater recharge was estimated using a separate model based on precipitation and potential evapotranspiration. Both hypothetical and GCM-predicted changes to the input parameters were used, resulting in higher, no different, and significantly lower recharge rates and groundwater elevations, depending on the climate scenario used.

Croley and Luukkonen (2003) investigated the impact of climate change on groundwater levels in Lansing, Michigan. The groundwater recharge rates were based on an empirical streamflow model which was calibrated using the results from two GCMs. The results of the study indicated that the simulated steady-state groundwater levels were generally predicted to increase or decrease due to climate change, depending on the GCM used.

Eckhardt and Ulbrich (2003) investigated the impact of climate change on groundwater recharge and streamflow in a small catchment in Germany. The input parameters in their hydrologic model were adjusted based on simulations from five different GCMs. The results of the study indicated that more precipitation will fall as rain in winter due to increased temperatures, resulting in higher recharge and streamflow in January and February. They also found that the increase in recharge from the snowmelt in March disappears, while recharge and streamflow were shown to be potentially reduced in the summer months.

Loaiciga (2003) studied a karst aquifer in south-central Texas and considered the impact of climate change not only on streambed recharge, but also on pumping rates (i.e. groundwater use). The impact of climate change on the streambed recharge was estimated using runoff scaling factors based on the ratio of historical and future streamflows predicted from linked general and regional climate models. The study concluded that the rise in groundwater use associated with predicted population growth would pose a higher threat to the aquifer than climate change.

Allen et al. (2004) used the Grand Forks aquifer, located in south-central British Columbia, Canada as a case study area for modeling the sensitivity of an aquifer to changes in recharge and river stage consistent with projected climate-change scenarios for the region. Results suggested that variations in recharge to the aquifer under the different climate-change scenarios, modeled under steady-state conditions, have a much smaller impact on the groundwater system than changes in river-stage elevation of the Kettle and Granby Rivers, which flow through the valley. All simulations showed relatively small changes in the overall configuration of the water table and general direction of groundwater flow. High-recharge and low-recharge simulations resulted in approximately a +0.05 m increase and a −0.025 m decrease, respectively, in water table elevations throughout the aquifer. Simulated changes in river-stage elevation, to reflect higher-than-peak flow levels (by 20 and 50%), resulted in average changes in the water-table elevation of 2.72 and 3.45 m, respectively. Simulated changes in river-stage elevation, to reflect lower-than-base flow levels (by 20 and 50%), resulted in average changes in the water-table elevation of -0.48 and -2.10 m, respectively. Current observed water table elevations in the valley are consistent with an average river-stage elevation (between current base flow and peak-flow stages).

Brouyere et al. (2004) developed an integrated hydrological model (MOHISE) in order to study the impact of climate change on the hydrological cycle in representative water basins in Belgium. This model considers most hydrological processes in a physically consistent way, more particularly groundwater flows which are modelled using a spatially distributed, finite-element approach. Thanks to this accurate numerical tool, after detailed calibration and validation, quantitative interpretations can be drawn from the groundwater model results. Considering IPCC climate change scenarios, the integrated approach was applied to evaluate the impact of climate change on the water cycle in the Geer basin in Belgium. The groundwater model is described in detail and results are discussed in terms of climate change impact on the evolution of groundwater levels and groundwater reserves. From the modelling application on the Geer basin, it appears that, on a pluri-annual basis, most tested scenarios predict a decrease in groundwater levels and reserves in relation to variations in climatic conditions. However, for this aquifer, the tested scenarios show no enhancement of the seasonal changes in groundwater levels.

Holman (2006) described an integrated approach to assess the regional impacts of climate and socio-economic change on groundwater recharge from East Anglia, UK. Many factors affect future groundwater recharge including changed precipitation and temperature regimes, coastal flooding, urbanization, woodland establishment, and changes in cropping and rotations. Important sources of uncertainty and shortcomings in recharge estimation were discussed in the light of the results. The uncertainty in, and importance of, socio-economic scenarios in exploring the consequences of unknown future changes were highlighted. Changes to soil properties are occurring over a range of time scales, such that the soils of the future may not have the same infiltration properties as existing soils. The potential implications involved in assuming unchanging soil properties were described.

Mall et al. (2006) examined the potential for sustainable development of surface water and groundwater resources within the constraints imposed by climate change and future research needs in India. In recent times, several studies around the globe show that climatic change is likely to impact significantly upon freshwater resources availability. In India, demand for water has already increased manifold over the years due to urbanization, agriculture expansion, increasing population, rapid industrialization and economic development. At present, changes in cropping pattern and land-use pattern, over-exploitation of water storage and changes in irrigation and drainage are modifying the hydrological cycle in many climate regions and river basins of India. An assessment of the availability of water resources in the context of future national requirements and expected impacts of climate change and its variability is critical for relevant national and regional long-term development strategies and sustainable development.

He concluded that the Indian region is highly sensitive to climate change. The elements/sectors currently at risk are likely to be highly vulnerable to climate change and variability. It is urgently required to intensify in-depth research work with the following objectives:

- Analyse recent experiences in climate variability and extreme events, and their impacts on regional water resources and groundwater availability.
- Study on changing patterns of rainfall, i.e. spatial and temporal variation and its impact on run-off and aquifer recharge pattern.
- Study sea-level rise due to increased run-off as projected due to glacial recession and increased rainfall.
- Sea-water intrusions into costal aquifers.
- Determine vulnerability of regional water resources to climate change and identify key risks and prioritize adaptation responses.
- Evaluate the efficacy of various adaptation strategies or coping mechanisms that may reduce vulnerability of the regional water resources.

It has been the endeavour of this study to summarize some important vulnerability issues associated with the present and potential future hydrological responses due to climate change and highlight those areas where further research is required. The National Environment Policy (2004) also advocated that anthropogenic climate changes have severe adverse impacts on India's precipitation patterns, ecosystems, agricultural potential, forests, water resources, coastal and marine resources. Large-scale planning would be clearly required for adaptation measures for climate change impacts, if catastrophic human misery is to be avoided.

Ranjan et al. (2006) evaluated the impacts of climate change on fresh groundwater resources specifically salinity intrusion in water resources stressed coastal aquifers. Their assessment used the Hadley Centre climate model, HadCM3 with high and low emission scenarios (SRES A2 and B2) for years 2000–2099. In both scenarios, the annual fresh groundwater resources losses indicated an increasing long-term trend in all stressed areas, except in the northern Africa/Sahara region. They also found that precipitation and temperature individually did not show good correlations with fresh groundwater loss. However, the relationship between the aridity index and fresh groundwater loss exhibited a strong negative correlation. They also discussed the impacts of loss of fresh groundwater resources on socio-economic activities, mainly population growth and per capita fresh groundwater resources.

Scibek and Allen (2006) developed a methodology for linking climate models and groundwater models to investigate future impacts of climate change on groundwater resources. An unconfined aquifer, situated near Grand Forks in south central British Columbia, Canada, was used to test the methodology. Climate change scenarios from the Canadian Global Coupled Model 1 (CGCM1) model runs were downscaled to local conditions using Statistical Downscaling Model (SDSM), and the change factors were extracted and applied in LARS-WG stochastic weather generator and then input to the recharge model. The recharge model simulated the direct recharge to the aquifer from infiltration of precipitation and consisted of spatially distributed recharge zones, represented in the Hydrologic Evaluation of Landfill Performance (HELP) hydrologic model linked to a geographic information system (GIS). A three-dimensional transient groundwater flow model, implemented in MODFLOW, was then used to simulate four climate scenarios in 1-year runs (1961–1999 present, 2010–2039, 2040–2069, and 2070-2099) and compare groundwater levels to present. The effect of spatial distribution of recharge on groundwater levels, compared to that of a single uniform recharge zone, is much larger than that of temporal variation in recharge, compared to a mean annual recharge representation. The predicted future climate for the Grand Forks area from the downscaled CGCM1 model will result in more recharge to the unconfined aquifer from spring to the summer season. However, the overall effect of recharge on the water balance is small because of dominant river-aquifer interactions and river water recharge.

Hsu et al. (2007) adopted a numerical modeling approach to investigate the response of the groundwater system to climate variability to effectively manage the groundwater resources of the Pingtung Plain. The Pingtung Plain is one of the most important groundwater-resource areas in southwestern Taiwan. The overexploitation of groundwater in the last two decades has led to serious deterioration in the quantity and quality of groundwater resources in this area. Furthermore, the manifestation of climate change tends to induce the instability of surface-water resources and strengthen the importance of the groundwater resources. Southwestern Taiwan in particular shows decreasing tendencies in both the annual amount of precipitation and annual precipitation days. A hydrogeological model was constructed based on the information from geology, hydrogeology, and geochemistry. Applying the linear regression model of precipitation to the next two decades, the modeling result shows that the lowering water level in the proximal fan raises an alarm regarding the decrease of available groundwater in the stress of climate change, and the enlargement of the low-groundwater-level area on the coast signals the deterioration of water quantity and quality in the future. Suitable strategies for water-resource management in response to hydrological impacts of future climatic change are imperative.

Jyrkama and Sykes (2007) presented a physically based methodology that can be used to characterize both the temporal and spatial effect of climate change on groundwater recharge. The method, based on the hydrologic model HELP3, can be used to estimate potential groundwater recharge at the regional scale with high spatial and temporal resolution. In this study, the method is used to simulate the past conditions, with 40 years of actual weather data, and future changes in the hydrologic cycle of the Grand River watershed. The impact of climate change is modelled by perturbing the model input parameters using predicted changes in the regions climate. The results of the study indicate that the overall rate of groundwater recharge is predicted to increase as a result of climate change. The higher intensity and frequency of precipitation will also contribute significantly to surface runoff, while global warming may result in increased evapotranspiration rates. Warmer winter temperatures will reduce the extent of ground frost and shift the spring melt from spring toward winter, allowing more water to infiltrate into the ground. While many previous climate change impact studies have focused on the temporal changes in groundwater recharge, results of this study suggest that the impacts can also have high spatial variability.

Toews (2007) modeled the impacts of future predicted climate change on groundwater recharge resources for the arid to semi-arid south Okanagan region, British Columbia. The hydrostratigraphy of the region consists of Pleistocene-aged glaciolacustrine silt overlain by glaciofluvial sand and gravel. Spatial recharge was modelled using available soil and climate data with the HELP 3.80D hydrology model. Climate change effects on recharge were investigated using stochastically-generated climate from three GCMs. Recharge is estimated to be ~45 mm/year, with minor increases expected with climate change. However, growing season and crop water demands will increase, posing additional stresses on water use in the region. A transient MODFLOW groundwater model simulates increases of water table in future time periods, which is largely driven by irrigation application increases. Spatial recharge was also used in a groundwater model to define capture zones around eight municipal water wells. These capture zones will be used for community planning.

Woldeamlak et al. (2007) modeled the effects of climate change on the groundwater systems in the Grote-Nete catchment, Belgium, covering an area of 525 km2, using wet (greenhouse), cold or NATCC (North Atlantic Thermohaline Circulation Change) and dry climate scenarios. Low, central and high estimates of temperature changes were adopted for wet scenarios. Seasonal and annual water balance components including groundwater recharge were simulated using the WetSpass model, while mean annual groundwater elevations and discharge were simulated with a steady-state MODFLOW groundwater model. WetSpass results for the wet scenarios showed that wet winters and drier summers are expected relative to the present situation. MODFLOW results for wet high scenario showed groundwater levels increase by as much as 79 cm, which could affect the distribution and species richness of meadows. Results obtained for cold scenarios depict drier winters and wetter summers relative to the present. The dry scenarios predict dry conditions for the whole year. There is no recharge during the summer, which is mainly attributed to high evapotranspiration rates by forests and low precipitation. Average annual groundwater levels drop by 0.5 m, with maximum of 3.1 m on the eastern part of the Campine Plateau. This could endanger aquatic ecosystem, shrubs, and crop production.

Carneiro et al. (2008) applied a density dependent numerical flow model (FEMWATER) to study the climate change impact in an unconfined shallow aquifer in the Mediterranean coast of Morocco. The stresses imposed to the model were derived from the IPCC emission scenarios and included recharge variations, rising sea level and advancing seashore. The simulations show that there will be a significant decline in the renewable freshwater resources and that salinity increases can be quite large but are limited to a restricted area.

Dragoni and Sukhija (2008) analysed the main methods for studying the relationships between climate change and groundwater, and presented the main areas in which hydrogeological research should focus in order to mitigate the likely impacts. There is a general consensus that climate change is an ongoing phenomenon. This will inevitably bring about numerous environmental problems, including alterations to the hydrological cycle, which is already heavily influenced by anthropogenic activity. The available climate scenarios indicate areas where rainfall may increase or diminish, but the final outcome with respect to man and environment will, generally, be detrimental. Groundwater will be vital to alleviate some of the worst drought situations.

Herrera-Pantoja and Hiscock (2008) outlined a methodology to quantify the effects of climate change on potential groundwater recharge (or hydrological excess water) for three locations in the north and south of Great Britain. Using results from a stochastic weather generator, actual evapotranspiration and potential groundwater recharge time-series for the historic baseline 1961-1990 and for a future 'high' greenhouse gas emissions scenario for the 2020s, 2050s and 2080s time periods were simulated for Coltishall in East Anglia, Gatwick in southeast England and Paisley in west Scotland. Under the 'high' gas emissions scenario, results showed a decrease of 20% in potential groundwater recharge for Coltishall, 40% for Gatwick and 7% for Paisley by the end of this century. The persistence of dry periods is shown to increase for the three sites during the 2050s and 2080s. Gatwick presents the driest conditions, Coltishall the largest variability of wet and dry periods and Paisley little variability. For Paisley, the main effect of climate change is evident during the dry season (April-September), when the potential amount of hydrological excess water decreases by 88% during the 2080s. Overall, it is concluded that future climate may present a decrease in potential groundwater recharge that will increase stress on local and regional groundwater resources that are already under ecosystem and water supply pressures.

Franssen (2009) indicated the limitations of the studies that address the impact of climate change on groundwater resources and suggested an improved approach. A general review, both from a groundwater hydrological and a climatological viewpoint, is given, oriented on the impact of climate change on groundwater resources. The impact of climate change on groundwater resources is not the subject of many studies in the scientific literature. Only rarely sophisticated downscaling techniques are applied to downscale estimated global circulation model (GCM) future precipitation series for a point or region of interest. Often it is not taken into account that different climate models calculate considerably different precipitation amounts (conceptual uncertainty). The joint downscaling of the meteorological variables that govern potential evapotranspiration (ET) is never done in the context of a study that assessed the impact of climate change on groundwater resources. It is desirable that actual ET is calculated in (groundwater) hydrological models on a physical basis, i.e. by coupling the energy and water balance at the Earth's surface.

Holman et al. (2009) indicated that groundwater resource estimates require the calculation of recharge using a daily time step. Within climate-change impact studies, this inevitably necessitates temporal downscaling of global or regional climate model outputs. This paper compares future estimates of potential groundwater recharge calculated using a daily soil-water balance model and climate-change weather time series derived using change factor (deterministic) and weather generator (stochastic) methods for Coltishall, UK. The uncertainty in the results for a given climate-change scenario arising from the choice of downscaling method is greater than the uncertainty due to the emissions scenario within a 30-year time slice. Robust estimates of the impact of climate change on groundwater resources require stochastic modelling of potential recharge, but this has implications for groundwater model runtimes. It is recommended that stochastic modelling of potential recharge is used in vulnerable or sensitive groundwater systems, and that the multiple recharge time series are sampled according to the distribution of contextually important time series variables, e.g. recharge drought severity and persistence (for water resource management) or high recharge years (for groundwater flooding). Such an approach will underpin an improved understanding of climate change impacts on sustainable groundwater resource management based on adaptive management and risk-based frameworks.

Shah (2009) reviewed the India's opportunities for mitigation and adaptation with reference to climate change and groundwater. For millennia, India used surface storage and gravity flow to water crops. During the last 40 years, however, India has witnessed a decline in gravity-flow irrigation and the rise of a booming 'water-scavenging' irrigation economy through millions of small, private tubewells. For India, groundwater has become at once critical and threatened.

Climate change will act as a force multiplier; it will enhance groundwater's criticality for drought-proofing agriculture and simultaneously multiply the threat to the resource. Groundwater pumping with electricity and diesel also accounts for an estimated 16–25 million mt of carbon emissions, 4–6% of India's total. From a climate change point of view, India's groundwater hotspots are western and peninsular India. These are critical for climate change mitigation as well as adaptation. To achieve both, India needs to make a transition from surface storage to 'managed aquifer storage' as the center pin of its water strategy with proactive demand- and supply-side management components. In doing this, India needs to learn intelligently from the experience of countries like Australia and the United States that have long experience in managed aquifer recharge.

Allen (2010) examined historical groundwater levels for selected observation wells in the south coastal region of British Columbia, Canada, to gain a better understanding of historical trends. Over a common period (1976-1999), negative trends in groundwater level dominate most records, and appear to be related to longer term negative regional trends in precipitation, although variable trends are evident at the shorter time periods used for this study. To explore potential consequences of varying recharge on groundwater quality, water chemistry data from selected monitoring wells on one island were examined. Chloride concentrations were observed to vary annually in one well by up to 4000 mg/L. Projections for future climate from one global climate model (CGCM1) were used as input to a recharge model to study the sensitivity of recharge to shifts in precipitation and temperature predicted for the region. The recharge model was driven by a stochastic daily weather series, calibrated to historic climate data. Daily weather series represent historic climate, and two future time periods (2020s) and (2050s). Simulated recharge increases progressively in the future using this particular global climate model; however, precipitation projections for this region of British Columbia are highly uncertain. Both positive and negative shifts in annual precipitation were predicted using a range of global climate models.

Allen et al. (2010) addressed variations in the prediction of recharge by comparing recharge simulated using climate data generated using a state-of-the-art downscaling method, TreeGen, with a range of global climate models (GCMs). The study site is the transnational Abbotsford-Sumas aquifer in coastal British Columbia, Canada and Washington State, USA, and is representative of a wet coastal climate. Sixty-four recharge zones were defined based on combinations of classed soil permeability, vadose zone permeability, and unsaturated zone depth (or depth to water table) mapped in the study area. One-dimensional recharge simulations were conducted for each recharge zone using the HELP hydrologic model, which simulates percolation through a vertical column. The HELP model is driven by mean daily temperature, daily precipitation, and daily solar radiation. For the historical recharge simulations, the climate data series was generated using the LARS-WG stochastic weather generator. Historical recharge was compared to recharge simulated using climate data series derived from the TreeGen downscaling model for three future time periods: 2020s (2010–2039), 2050s (2040–2069), and 2080s (2070–2099) for each of four GCMs (CGCM3.1, ECHAM5, PCM1, and CM2.1). Recharge results are compared on an annual basis for the entire aquifer area. Both increases and decreases relative to historical recharge are simulated depending on time period and model. By the 2080s, the range of model predictions spans −10.5% to +23.2% relative to historical recharge. This variability in recharge predictions suggests that the seasonal performance of the downscaling tool is important and that a range of GCMs should be considered for water management planning.

Crosbie et al. (2010) presented a methodology for assessing the average changes in groundwater recharge under a future climate. The method is applied to the 1,060,000 km^2 Murray-Darling Basin (MDB) in Australia. Climate sequences were developed based upon three scenarios for a 2030 climate relative to a 1990 climate from the outputs of 15 global climate models. Dryland diffuse groundwater recharge was modelled in WAVES using these 45 climate scenarios and fitted to a Pearson Type III probability distribution to condense the 45 scenarios down to three: a wet future, a median future and a dry future. The use of a probability distribution allowed the significance of any change in recharge to be assessed. This study found that for the median future, climate recharge is projected to increase on average by 5% across the MDB but this is not spatially uniform. In the wet and dry future scenarios the recharge is projected to increase by 32% and decrease by 12% on average across the MDB, respectively. The differences between the climate sequences generated by the 15 different global climate models makes it difficult to project the direction of the change in recharge for a 2030 climate, let alone the magnitude.

Dams et al. (2010) presented a methodology to predict the potential impact of climate change on quantitative groundwater characteristics determining GWDTEs (Groundwater Dependent Terrestrial Ecosystems). The developed methodology includes coupling a distributed hydrological model (WetSpa) with a transient groundwater flow model (MODFLOW) and is tested for the Kleine Nete basin, Belgium. Because the occurrence of phreatophytes is strongly determined by the dynamic properties of the groundwater system, a groundwater flow model with a high temporal and spatial distribution was developed using MODFLOW. The groundwater recharge and river heads are estimated with the WetSpa model using a daily time step to incorporate the impact of changes in rainfall intensity. Potential future hydrological changes are calculated by comparing the hydrological state corresponding to 1960-1991 with future scenarios developed for 2070-2101.

Since the uncertainty in the prediction of the future climate components such as potential evapotranspiration (PET) and precipitation is still high, an ensemble of 28 climate scenarios were chosen from the PRUDENCE database. For each of these scenarios, recharge, river stage, groundwater head and groundwater flow are estimated for 32 years with half monthly time steps. Comparison of the original measured PET with future PET shows that the PET during summer rises in all future scenarios with about 1 mm per day. For winter conditions the scenarios predict little change in PET. Future precipitation shows an increase in precipitation during winter and a decrease during summer. Future groundwater recharge decreases on average with 20 mm per year, the highest decreases are simulated from July until September. Average groundwater heads indicate an average decrease of 7 cm. Groundwater levels in interfluves generally show decreases up to 30 cm. The mean lowest groundwater level decreases on average with 6 cm, while the mean highest groundwater level decreases about 3 cm. On average, the groundwater discharge reduces with 4%, from 5 to 4.8 m³/s. GWDTEs that currently receive a low groundwater discharge, are likely to disappear due to future climate changes.

McCallum et al. (2010) used a sensitivity analysis of climate variables using a modified version of WAVES, a soil-vegetation-atmosphere-transfer model (unsaturated zone), to determine the importance of each climate variable in the change in groundwater recharge for three points in Australia. This study found that change in recharge is most sensitive to change in rainfall. Increases in temperature and changes in rainfall intensity also led to significant changes in recharge. Although not as significant as other climate variables, some changes in recharge were observed due to changes in solar radiation and carbon dioxide concentration. When these variables were altered simultaneously, changes in recharge appeared to be closely related to changes in rainfall; however, in nearly all cases, recharge was greater than would have been predicted if only rainfall had been considered. These findings have implications for how recharge is projected to change due to climate change.

Okkonen et al. (2010) presented a literature review of the impacts of anticipated climate change on unconfined aquifers, along with a conceptual framework for evaluating the complex responses of surface and subsurface hydrology to climate variables in cold regions. The framework offers a way to conceptualize how changes in one component of the system may impact another by delineating the relationships among climate drivers, hydrological responses, and groundwater responses in a straight-forward manner. The model is elaborated in the context of shallow unconfined aquifers in the boreal environment of Finland. In cold conditions, climate change is expected to reduce snow cover and soil frost and increase winter floods. The annual surface water level maximum will occur earlier in spring, and water levels will decrease in summer due to higher evapotranspiration rates. The maximum recharge and groundwater level are expected to occur earlier in the year. Lower groundwater levels are expected in summer due to higher evapotranspiration rates. The flow regimes between shallow unconfined aquifers and surface water may change, affecting water quantity and quality in the surface and groundwater systems.

Oude Essink et al. (2010) focussed on a coastal groundwater system that is already threatened by a relatively high seawater level: the low-lying Dutch Delta. Nearly one third of the Netherlands lies below mean sea level, and the land surface is still subsiding up to 1 m per century. This densely populated delta region, where fresh groundwater resources are used intensively for domestic, agricultural, and industrial purposes, can serve as a laboratory case for other low-lying delta areas throughout the world. Their findings on hydrogeological effects can be scaled up since the problems the Dutch face now will very likely be the problems encountered in other delta areas in the future. They calculated the possible impacts of future sea level rise, land subsidence, changes in recharge, autonomous salinization, and the effects of two mitigation counter-measures with a three-dimensional numerical model for variable density groundwater flow and coupled solute transport (MOCDENS3D). They considered the effects on hydraulic heads, seepage fluxes, salt loads to surface waters, and changes in fresh groundwater resources as a function of time and for seven scenarios.

Their numerical modeling results show that the impact of sea level rise is limited to areas within 10 km of the coastline and main rivers because the increased head in the groundwater system at the coast can easily be produced though the highly permeable Holocene confining layer. Along the southwest coast of the Netherlands, salt loads will double in some parts of the deep and large polders by the year 2100 A.D. due to sea level rise. More inland, ongoing land subsidence will cause hydraulic heads and phreatic water levels to drop, which may result in damage to dikes, infrastructure, and urban areas. In the deep polders more inland, autonomous upconing of deeper and more saline groundwater will be responsible for increasing salt loads. The future increase of salt loads will cause salinization of surface waters and shallow groundwater and put the total volumes of fresh groundwater volumes for drinking water supply, agricultural purposes, industry, and ecosystems under pressure.

Payne (2010) observed that sea-level rise and changes in precipitation patterns may contribute to the occurrence and affect the rate of saltwater contamination in the Hilton Head Island, South Carolina area. To address the effects of climate change on saltwater intrusion, a three-dimensional, finite-element, variable-density, solute-transport model (SUTRA 2.1) was developed to simulate different rates of sea-level rise and variation in onshore freshwater recharge. Model simulation showed that the greatest effect on the existing saltwater plume occurred from reducing recharge, suggesting recharge may be a more important consideration in saltwater intrusion management than estimated rates of sea-level rise. Saltwater intrusion management would benefit from improved constraints on recharge rates by using model-independent, local precipitation and evapotranspiration data, and improving estimates of confining unit hydraulic properties.

Rozell and Wong (2010) investigated the effects of climate change on Shelter Island, New York State (USA), a small sandy island, using a variable-density transient groundwater flow model (SEAWAT). Predictions for changes in precipitation and sea-level rise over the next century from the Intergovernmental Panel on Climate Change 2007 report were used to create two future climate scenarios. In the scenario most favorable to fresh groundwater retention, consisting of a 15% precipitation increase and 0.18-m sea-level rise, the result was a 23-m seaward movement of the freshwater/ salt-water interface, a 0.27-m water-table rise, and a 3% increase in the fresh-water lens volume. In the scenario supposedly least favorable to groundwater retention, consisting of a 2% precipitation decrease and 0.61-m sea-level rise, the result was a 16-m landward movement of the fresh-water/salt-water interface, a 0.59-m watertable rise, and a 1% increase in lens volume. The unexpected groundwater-volume increase under unfavorable climate change conditions was best explained by a clay layer under the island that restricts the maximum depth of the aquifer and allows for an increase in freshwater lens volume when the water table rises.

Vandenbohede and Lebbe (2010) evaluated the effects of sea level rise and future recharge changes on the coastal aquifer of the western Belgian coastal plain with a 3D density dependent groundwater flow model (MOCDENS3D). The area is characterised by a wide dune belt. Sea level rise results in a landward enlargement of the fresh water lens under the dunes and an increase of flow towards the dune-polder transition's drainage system. Recharge increase results also in an enlargement of the dune's fresh water lens and an increase of the amount of water which must be evacuated by the polder's drainage system. Recharge decrease has the reverse effect.

Zhou et al. (2010) reported that climate change affects not only water resources but also water demand for irrigation. A large proportion of the world's agriculture depends on groundwater, especially in arid and semi-arid regions. In several regions, aquifer resources face depletion. Groundwater recharge has been viewed as a by-product of irrigation return flow, and with climate change, aquifer storage of such flow will be vital. A general review, for a broadbased audience, is given of work on global warming and groundwater resources, summarizing the methods used to analyze the climate change scenarios and the influence of these predicted changes on groundwater resources around the world (especially the impact on regional groundwater resources and irrigation requirements). Future challenges of adapting to climate change are also discussed. Such challenges include water-resources depletion, increasing irrigation demand, reduced crop yield, and groundwater salinization. The adaptation to and mitigation of these effects is also reported, including useful information for water-resources managers and the development of sustainable groundwater irrigation methods. Rescheduling irrigation according to the season, coordinating the groundwater resources and irrigation demand, developing more accurate and complete modeling prediction methods, and managing the irrigation facilities in different ways would all be considered, based on the particular cases.

Barthel (2011) presented an integrated approach for assessing the availability of groundwater under conditions of 'global-change'. The approach is embedded in the DANUBIA system developed by the interdisciplinary GLOWA-Danube Project to simulate the interaction of natural and socio-economic processes within the Upper Danube Catchment (UDC, 77,000 km^2 and located in parts of Germany, Austria, Switzerland and Italy). The approach enables the quantitative assessment of groundwater bodies (zones), which are delineated by intersecting surface watersheds, regional aquifers, and geomorphologic regions. The individual hydrogeological and geometrical characteristics of these zones are accounted for by defining characteristic response times and weights to describe the relative significance of changes in variables (recharge, groundwater level, groundwater discharge, river discharge) associated with different states. These changes, in each zone, are converted into indices (*GroundwaterQuantityFlags*). The motivation and particularities of regional-scale groundwater assessment and the background of GLOWA-Danube are described, along with a description of the developed methodology. The approach was applied to the UDC, where several different climate scenarios (2011–2060) were evaluated. A selection of results is presented to demonstrate the potential of the methodology. The approach was inspired by the European Water Framework Directive, yet it has a stronger focus on the evaluation of global-change impacts.

Yihdego and Webb (2011) carried out modeling of bore hydrographs to determine the impact of climate and land-use change in a temperate subhumid region of southeastern Australia. To determine the relative impact of climate and human intervention on groundwater elevations in western Victoria, southeast Australia, bore hydrograph fluctuations in three aquifers were modelled using a transfer function noise model (PIRFICT) and an auto-regressive model (HARTT), which give generally comparable results. Most of the groundwater-level fluctuations (>90%) are explained by climatic variation, particularly rainfall. The overall non-climate-related trend in groundwater level is downward and small but statistically significant (−0.04 to −0.066 m/yr), and is probably due to the widespread replacement of grazing land by wheat and canola cultivation, as these crops use more water than pasture. A large non-climate-related trend (−0.30 m/yr) for bores in an irrigation area is mainly related to groundwater extraction. The response time of the system is rapid (only 4.85 years on average), much faster than previously estimated. Rates of groundwater flow are much slower; groundwater ages are up to ~35,000 years. Response times effectively represent the time for the system to move to a new state of hydrologic equilibrium; this prediction of the time scale of the impacts of land-use change on groundwater resources will allow the development of better strategies for groundwater management.

Crosbie et al. (2012) investigated episodic recharge with reference to climate change in the Murray-Darling Basin, Australia. In semi-arid areas, episodic recharge can form a significant part of overall recharge, dependant upon infrequent rainfall events. With climate change projections suggesting changes in future rainfall magnitude and intensity, groundwater recharge in semi-arid areas is likely to be affected disproportionately by climate change. This study sought to investigate projected changes in episodic recharge in arid areas of the Murray-Darling Basin, Australia, using three global warming scenarios from 15 different global climate models (GCMs) for a 2030 climate. Two metrics were used to investigate episodic recharge: at the annual scale the coefficient of variation was used, and at the daily scale the proportion of recharge in the highest 1% of daily recharge. The metrics were proportional to each other but were inconclusive as to whether episodic recharge was to increase or decrease in this environment; this is not a surprising result considering the spread in recharge projections from the 45 scenarios. The results showed that the change in the low probability of exceedance rainfall events was a better predictor of the change in total recharge than the change in total rainfall, which has implications for the selection of GCMs used in impact studies and the way GCM results are downscaled.

Neukum and Azzam (2012) investigated the impact of climate change on groundwater recharge in a small catchment in the Black Forest, Germany. Temporal and spatial changes of the hydrological cycle are the consequences of climate variations. In addition to changes in surface runoff with possible floods and droughts, climate variations may affect groundwater through alteration of groundwater recharge with consequences for future water management. This study investigates the impact of climate change, according to the Special Report on Emission Scenarios (SRES) A1B, A2 and B1, on groundwater recharge in the catchment area of a fissured aquifer in the Black Forest, Germany, which has sparse groundwater data. The study uses a water-balance model considering a conceptual approach for groundwater-surface water exchange. River discharge data are used for model calibration and validation. The results show temporal and spatial changes in groundwater recharge. Groundwater recharge is progressively reduced for summer during the twenty-first century. The annual sum of groundwater recharge is affected negatively for scenarios A1B and A2. On average, groundwater recharge during the twenty-first century is reduced mainly for the lower parts of the valley and increased for the upper parts of the valley and the crests. The reduced storage of water as snow during winter due to projected higher air temperatures causes an important relative increase in rainfall and, therefore, higher groundwater recharge and river discharge.

Raposo et al. (2013) assessed the impact of future climate change on groundwater recharge in Galicia-Costa, Spain. Climate change can impact the hydrological processes of a watershed and may result in problems with future water supply for large sections of the population. Results from the FP5 PRUDENCE project suggest significant changes in temperature and precipitation over Europe. In this study, the Soil and Water Assessment Tool (SWAT) model was used to assess the potential impacts of climate change on groundwater recharge in the hydrological district of Galicia-Costa, Spain. Climate projections from two general circulation models and eight different regional climate models were used for the assessment and two climate-change scenarios were evaluated. Calibration and validation of the model were performed using a daily time-step in four representative catchments in the district. The effects on modeled mean annual groundwater recharge are small, partly due to the greater stomatal efficiency of plants in response to increased CO_2 concentration. However, climate change strongly influences the temporal variability of modeled groundwater recharge. Recharge may concentrate in the winter season and dramatically decrease in the summer–autumn season. As a result, the dry-season duration may be increased on average by almost 30 % for the A2 emission scenario, exacerbating the current problems in water supply.

Lapworth et al. (2013) estimated residence times of shallow groundwater in West Africa. Although shallow groundwater (<50 mbgl) sustains the vast majority of improved drinking-water supplies in rural Africa, there is little information on how resilient this resource may be to future changes in climate. This study presents results of a groundwater survey using stable isotopes, CFCs, SF_6, and ^3H across different climatic zones (annual rainfall 400–2,000 mm/year) in West Africa. The purpose was to quantify the residence times of shallow groundwaters in sedimentary and basement aquifers, and investigate the relationship between groundwater resources and climate. Stable-isotope results indicate that most shallow groundwaters are recharged rapidly following rainfall, showing little evidence of evaporation prior to recharge. Chloride mass-balance results indicate that within the arid areas (<400 mm annual rainfall) there is recharge of up to 20 mm/year. Age tracers show that most groundwaters have mean residence times (MRTs) of 32–65 years, with comparable MRTs in the different climate zones. Similar MRTs measured in both the sedimentary and basement aquifers suggest similar hydraulic diffusivity and significant groundwater storage within the shallow basement. This suggests there is considerable resilience to short-term inter-annual variation in rainfall and recharge, and rural groundwater resources are likely to sustain diffuse, low volume abstraction.

Mollema and Antonellini (2013) investigated seasonal variation in natural recharge of coastal aquifers. Many coastal zones around the world have irregular precipitation throughout the year. This results in discontinuous natural recharge of coastal aquifers, which affects the size of freshwater lenses present in sandy deposits. Temperature data for the period 1960–1990 from LocClim (local climate estimator) and those obtained from the Intergovernmental Panel on Climate Change (IPCC) SRES A1b scenario for 2070–2100, have been used to calculate the potential evapotranspiration with the Thornthwaite method. Potential recharge (difference between precipitation and potential evapotranspiration) was defined at 12 locations: Ameland (The Netherlands), Auckland and Wellington (New Zealand); Hong Kong (China); Ravenna (Italy), Mekong (Vietnam), Mumbai (India), New Jersey (USA), Nile Delta (Egypt), Kobe and Tokyo (Japan), and Singapore. The influence of variable/discontinuous recharge on the size of freshwater lenses was simulated with the SEAWAT model. The discrepancy between models with continuous and with discontinuous recharge is relatively small in areas where the total annual recharge is low (258–616 mm/year); but in places with Monsoon-dominated climate (e.g. Mumbai, with recharge up to 1,686 mm/year), the difference in freshwater-lens thickness between the discontinuous and the continuous model is larger (up to 5 m) and thus important to consider in numerical models that estimate freshwater availability.

These studies are still at infancy and more data, in terms of field information are to be generated. This will also facilitate appropriate validation of the simulation for the present scenarios. In summary, climate change is likely to have an impact on future recharge rates and hence on the underlying groundwater resources. The impact may not necessarily be a negative one, as evidenced by some of the investigations. Quantifying the impact is difficult, however, and is subject to uncertainties present in the future climate predictions. Simulations based on general circulation models (GCMs) have yielded mixed and conflicting results, raising questions about their reliability in predicting future hydrologic conditions. However, it is clear that the global warming threat is real and the consequences of climate change phenomena are many and alarming.

Methodology to Assess the Impact of Climate Change on Groundwater Resources

The potential impacts of climate change on water resources have long been recognized although there has been comparatively little research relating to groundwater. The principle focus of climate change research with regard to groundwater has been on quantifying the likely direct impacts of changing precipitation and temperature patterns. Such studies have used a range of modelling techniques such as soil water balance models, empirical models, conceptual models and more complex distributed models, but all have derived changes in groundwater recharge assuming parameters other than precipitation and temperature remaining constant.

There are two main parameters that could have a significant impact on groundwater levels: recharge and river stage/discharge. To assess the impact on the groundwater system to changes in these two parameters, it is necessary to have a calibrated flow model and to conduct a sensitivity analysis by varying these two parameters and calculating changes to the water balance (e.g., differences in water levels). The research objectives can be:

> To develop a conceptual model of the hydrogeology of the study region;

> To investigate how regional and local weather events affect recharge;

> To determine potential impacts of climate change on recharge for the study area, and to assess the sensitivity of the results to different global climate models;

> To develop and calibrate a regional-scale three-dimensional groundwater flow model of the region and to use that model to assess the impacts of climate change on groundwater resources; and

> To develop and calibrate a local-scale three-dimensional groundwater flow model, and to undertake a well capture zone analysis for the local community water supply wells for the region.

The methodology consists of three main steps. To begin with, climate scenarios can be formulated for the future years such as 2050 and 2100. This is done by assigning percentage or value changes of climatic variables on a seasonal and/or annual basis only for the future years relative to the present year. Secondly, based on these scenarios and present situation, seasonal and annual recharge, evapotranspiration and runoff are simulated with the WHI UnSat Suite (HELP module for recharge) and/or WetSpass model. Finally, the annual recharge outputs from WHI UnSat Suite or WetSpass model are used to simulate groundwater system conditions using steady-state MODFLOW model setups for the present condition and for the future years.

The main tasks that are involved in such a study are:

1. Describe hydrogeology of the study area.

2. Undertake a statistical analysis to separate climate into regional and local events and determine the role of each in contributing to groundwater recharge.

3. Analyze climate data from weather stations and modelled GCM, and build future predicted climate change datasets with temperature, precipitation and solar radiation variables.

4. Define methodology for estimating changes to recharge in the model under both current climate conditions and for the range of climate-change scenarios for the study area.

5. Use of a computer code (such as WHI UnSat Suite or WetSpass) to estimate recharge based on available precipitation and temperature records and anticipated changes to these parameters.

Recharge estimation by WHI UnSat Suite

UnSat Suite contains the subprogram, Visual HELP, which contains a more user-friendly interface for the program HELP that is approved by the United States Environmental Protection Agency (US EPA) for designing landfills. Visual HELP enables the modeler to generate estimates of recharge using a weather generator and the properties of the aquifer column.

Recharge estimation by WetSpass

WetSpass is a quasi physically distributed seasonal-water balance model, which takes into account detailed soil, land-use, slope, groundwater depth, and hydro-climatological distributed maps with associated parameter tables for estimating groundwater recharge. The model uses seasonal (summer and winter) geographical information systems (GIS) input grids of the mentioned inputs to estimate annual and seasonal groundwater recharge values.

6. Quantify the spatially distributed recharge rates using the climate data and spatial soil survey data.

7. Development and calibration of a three-dimensional regional-scale groundwater flow model (such as Visual MODFLOW). Since one of the inputs required for WetSpass is the groundwater depth data, which is predicted with the MODFLOW model, an interface may be developed in an ArcView GIS platform to couple the two models, facilitating exchange of data between the two models. The coupled WetSpass-MODFLOW model is run for the present situation and for each of the climate change scenarios on an annual basis.

8. Simulate groundwater flow using each recharge data set and evaluate the changes in groundwater flow and levels through time.

9. Undertake sensitivity analysis of the groundwater flow model.

10. Develop a local scale groundwater model for the specific study area and conduct a well capture zone analysis.

A typical flow chart for various aspects of such a study is shown in Figure 2. The figure shows the connection from the climate analysis, to recharge simulation, and finally to a groundwater model. Recharge is applied to a three-dimensional groundwater flow model, which is calibrated to historical water levels. Transient simulations are undertaken to investigate the temporal response of the aquifer system to historic and future climate periods.

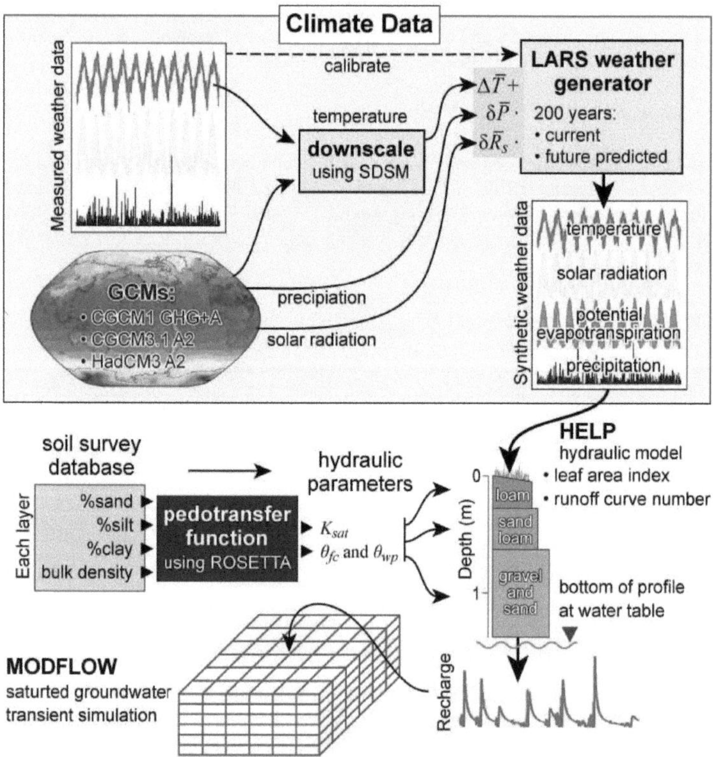

Figure 2: Flow Chart of Tasks (Toews et al., 2007)

Tasks in the upper part of the chart assemble several climate data sets for current and future predicted conditions, which are used to simulate recharge using HELP module of WHI UnSat Suite. The soil layers are parameterized using a pedotransfer function program, which utilizes detailed soil survey measurements. Mapped monthly recharge from HELP is then used in a three-dimensional MODFLOW model to simulate transient saturated groundwater flow.

Concluding Remarks

> Precipitation is commonly downscaled in climate change impact studies; however, the reliability of the downscaled result is often poor or unreliable, as there is often little correlation between the predictors and the predictands. A poor correlation is often attributed to mesoscale processes occurring at the site-scale that are not represented in regional models due to their representative spatial and temporal sizes in comparison to larger-scale regional precipitation. Mesoscale precipitation processes generally occur in the summer season in the form of convective clouds, which are a result of local-scale evapotranspiration from elevated temperatures and solar radiation magnitudes. As a result, global-scale models may underestimate the summer precipitation measured at a site.

> Although climate change has been widely recognized, research on the impacts of climate change on the groundwater system is relatively limited. The reasons may be that long historical data are required to analyze the characteristics of climate change. These data are not always available. Also, the driving forces that cause such changes are yet unclear. The climatic abnormality may occur frequently and last for a period of time. Even if the required data exist, uncertainty is embedded in model parameters, structure and driving force of the hydrological cycle. Predicting the long-term effect of a dynamic system is very difficult because of limitations inherent in the models, and the unpredictability of the forces that drive the earth. A physically based model of a groundwater system under possible climate change based on available data is very important to prevent the deterioration of regional water-resource problems in the future. Although uncertainties are inevitable, new response strategies in water resource management based on the model may be useful.

> The investigation of the relationship between climate change and loss of fresh groundwater resources is important for understanding the characteristics of the different regions. The impact of future climatic change may be felt more severely in developing countries such as India, whose economy is largely dependent on agriculture and is already under stress due to current population increase and associated demands for energy, freshwater and food. In spite of the uncertainties about the precise magnitude of climate change and its possible impacts, particularly on regional scales, measures must be taken to anticipate, prevent or minimize the causes of climate change and mitigate its adverse effects.

> Groundwater recharge is influenced not only by hydrologic processes, but also by the physical characteristics of the land surface and soil profile. Many climate change studies

have focused on modelling the temporal changes in the hydrologic processes and ignored the spatial variability of physical properties across the study area. While knowing the average change in recharge and groundwater levels over time is important, these changes will not occur equally over a regional catchment or watershed. Long-term water resource planning requires both spatial and temporal information on groundwater recharge in order to properly manage not only water use and exploitation, but also land use allocation and development. Studies concerned with climate change should therefore also consider the spatial change in groundwater recharge rates.

> If the likely consequences of future changes of groundwater recharge, resulting from both climate and socio-economic change, are to be assessed, hydrogeologists must increasingly work with researchers from other disciplines, such as socio-economists, agricultural modelers and soil scientists.

References

Allen, Diana M. (2010), Historical trends and future projections of groundwater levels and recharge in coastal British Columbia, Canada, SWIM21 - 21st Salt Water Intrusion Meeting, 21-26 June 2010, Azores, Portugal, pp. 267-270.

Allen, D. M., Mackie, D. C. and Wei, M. (2004), Groundwater and climate change: a sensitivity analysis for the Grand Forks aquifer, southern British Columbia, Canada, Hydrogeology Journal, Vol. 12, pp. 270–290.

Allen, D. M., Cannon, A. J., Toews, M. W. and Scibek, J. (2010), Variability in simulated recharge using different GCMs, Water Resources Research, Vol. 46.

Barthel, Roland (2011), An indicator approach to assessing and predicting the quantitative state of groundwater bodies on the regional scale with a special focus on the impacts of climate change, Hydrogeology Journal, Volume 19, Issue 3, pp. 525-546.

Bouraoui, F., Vachaud, G., Li, L. Z. X., Le Treut, H. and Chen, T. (1999), Evaluation of the impact of climate changes on water storage and groundwater recharge at the watershed scale, Climate Dynamics, Vol. 15, pp. 153-161.

Brouyere, S., Carabin, G. and Dassargues, A. (2004), Climate change impacts on groundwater resources: modelled deficits in a chalky aquifer, Geer basin, Belgium, Hydrogeology Journal, Vol. 12, pp. 123-134.

Carneiro, Júlio F., Boughriba, M., Correia, A., Zarhloule, Y., Rimi, A. and EL Houadi, B. (2008), Climate Change Impact in a Shallow Coastal Mediterranean Aquifer, at Saïdia, Morocco, 20th Salt Water Intrusion Meeting, 23-27 June 2008, Naples, Florida, USA, pp. 30-33.

Central Ground Water Board (2002), Master Plan for Artificial Recharge to Groundwater in India, Central Ground Water Board, New Delhi, February 2002, p. 115.

Central Ground Water Board (2011), Dynamic Ground Water Resources of India (as on 31 March 2009), Ministry of Water Resources, Government of India, November 2011, 225p.

Crosbie, Russell S., McCallum, James L., Walker, Glen R. and Chiew, Francis H. S. (2012), Episodic recharge and climate change in the Murray-Darling Basin, Australia, Hydrogeology Journal, Volume 20, Issue 2, pp. 245-261.

Chadha, D. K. and Sharma, S. K. (2000), Groundwater management in India issues and options, Workshop on Past Achievements and Future Strategies, Central Groundwater Authority, New Delhi, 14 January 2000.

Climate Change & its Impact on Indian Water Resources: Assessment, Adaptation & Mitigation, Base Paper by National Institute of Hydrology, Roorkee.

Croley, T. E. and Luukkonen, C. L. (2003), Potential effects of climate change on ground water in Lansing, Michigan, Journal of the American Water Resources Association, Vol. 39 (1), pp. 149-163.

Crosbie, Russell S., McCallum, James L., Walker, Glen R. and Chiew, Francis H. S. (2010), Modelling climate-change impacts on groundwater recharge in the Murray-Darling Basin, Australia, Hydrogeology Journal, Vol. 18, pp. 1639–1656.

Dams, Jef, Salvadore, Elga and Batelaan, Okke (2010), Predicting Impact of Climate Change on Groundwater Dependent Ecosystems, 2nd International Interdisciplinary Conference on Predictions for Hydrology, Ecology and Water Resources Management: Changes and Hazards caused by Direct Human Interventions and Climate Change, 20-23 September 2010, Prague, Czech Republic.

Das, P. K. and Radhakrishnan, M. (1991), Proc. Indian Acad. Sci. (Earth Planet. Sci.), Vol. 100, pp. 177–194.

Dragoni, W. and Sukhija, B. S. (2008), Climate change and groundwater: a short review, Geological Society, London, Special Publications 2008; v. 288; p. 1-12.

Eckhardt, K. and Ulbrich, U. (2003), Potential impacts of climate change on groundwater recharge and streamflow in a central European low mountain range, Journal of Hydrology, Vol. 284, pp. 244-252.

Franssen, Harrie-Jan Hendricks (2009), The impact of climate change on groundwater resources, International Journal of Climate Change Strategies and Management, Vol. 1, Issue 3, pp. 241-254.

Ghosh Bobba, A. (2002), Numerical modelling of salt-water intrusion due to human activities and sea-level change in the Godavari Delta, India, Hydrological Sciences Journal, Vol. 47(S), August 2002, pp. 67-80.

Herrera-Pantoja, M. and Hiscock, K. M. (2008), The effects of climate change on potential groundwater recharge in Great Britain, Hydrological Processes, Vol. 22, Issue 1, January 2008, pp. 73-86.

Holman, I. P. (2006), Climate change impacts on groundwater recharge - uncertainty, shortcomings, and the way forward?, Hydrogeology Journal, Vol. 14, pp. 637–647.

Holman, I. P., Tascone, D. and Hess, T. M. (2009), A comparison of stochastic and deterministic downscaling methods for modelling potential groundwater recharge under climate change in East Anglia, UK: implications for groundwater resource management, Hydrogeology Journal, Vol. 17, pp. 1629–1641.

Hsu, Kuo-Chin, Wang, Chung-Ho, Cheu, Kuan-Chih, Chen, Chien-Tai and Ma, Kai-Wei (2007), Climate-induced hydrological impacts on the groundwater system of the Pingtung Plain, Taiwan, Vol. 15, Number 5, August 2007, pp. 903-913.

IPCC (1995), Report of Working Group I of the Intergovernmental Panel on Climate Change, World Meteorological Organization, United Nations Environment Program, Geneva.

IPCC (1997), The regional impact of climate change: an assessment of vulnerability: A Special Report of IPCC working group II, Watson, R. T., Zinyowera, M. C., Moss, R. H., Dokken, D. J. (Eds), http://www.grida.no/climate/ipcc/regional/index.htm, Cited 19 November 2006.

IPCC (2000), Emissions Scenarios: A Special Report of Working Group II of the Intergovernmental Panel on Climate Change, In: Nakicenovic, N., Swart, R. (Eds.), Cambridge Univ. Press, Cambridge, UK.

IPCC (2001), In: Houghton, J. T., Ding, Y., Griggs, D. J., Noguer, M., van der Linden, P. J., Dai, X., Maskell, K., Johnson, C.A. (Eds.), Climate Change 2001: The Scientific Basis, Contributions of Working Group 1 to the Third Assessment Report of the Intergovernmental Panel on Climate Change, Cambridge University Press, Cambridge, UK.

IPCC (2001), In: James J. McCarthy, Osvaldo F. Canziani, Neil A. Leary, David J. Dokken, Kasey S. White (Eds.), Climate Change 2001: Impacts, Adaptation, and Vulnerability, Contribution of Working Group II to the Third Assessment Report of the Intergovernmental Panel on Climate Change, Cambridge University Press, Cambridge, UK.

IPCC (2007), In: Solomon, S., Qin, D., Manning, M., Chen, Z., Marquis, M., Averyt, K. B., Tignor, M., Miller, H. L. (eds.), 2007. Climate Change 2007: The Physical Science Basis. Contribution of Working Group I to the Fourth Assessment Report of the Intergovernmental Panel on Climate Change, Cambridge University Press, Cambridge, UK and New York, NY, USA, 966 p.

Jyrkama, Mikko I. and Sykes, Jon F. (2007), The impact of climate change on spatially varying groundwater recharge in the grand river watershed (Ontario), Journal of Hydrology, Vol. 338, pp. 237–250.

Kirshen, P. H. (2002), Potential impacts of global warming on groundwater in Eastern Massachusetts, Journal of Water Resources Planning and Management, Vol. 128 (3), pp. 216-226.

Kumar, C. P. and Surjeet Singh (2011), Impact of Climate Change on Water Resources and National Water Mission of India, Proceedings of National Seminar on Global Warming and its Impact on Water Resources, 14 January 2011, Kolkata, pp. 1-9 (Volume 1).

Kumar, C. P. and Surjeet Singh (2012), Impact of Climate Change on Groundwater Resources – Status of Research Studies, Proceedings of 3rd International Conference on Climate Change and Sustainable Management of Natural Resources, 5-7 February 2012, Gwalior.

Kumar, C. P. (2012), Assessing the Impact of Climate Change on Groundwater Resources, India Water Week 2012 - Water, Energy and Food Security: Call for Solutions, 10-14 April 2012, New Delhi.

Kumar C. P. (2012), Assessment of Impact of Climate Change on Groundwater Resources, Water and Energy International, ISSN: 0974-4711, Vol. 69, Number 8, August 2012, pp. 25-31.

Kumar, C. P. (2012), Climate Change and Its Impact on Groundwater Resources, Research Inventy: International Journal of Engineering and Science, ISSN: 2278-4721, Volume 1, Issue 5 October 2012, pp. 43-60.

Lapworth, D. J., MacDonald, A. M., Tijani, M. N., Darling, W. G., Gooddy, D. C., Bonsor, H. C. and Araguás-Araguás, L. J. (2013), Residence times of shallow groundwater in West Africa: implications for hydrogeology and resilience to future changes in climate, Hydrogeology Journal, Volume 21, Issue 3, pp. 673-686.

Loaiciga, H. A. (2003), Climate change and ground water, Annals of the Association of American Geographers, Vol. 93 (1), pp. 30–41.

Mall, R. K., Gupta, Akhilesh, Singh, Ranjeet, Singh, R. S. and Rathore, L. S. (2006), Water resources and climate change: An Indian perspective, Current Science, Vol. 90, No. 12, 25 June 2006.

McCallum, J. L., Crosbie, R. S., Walker, G. R. and Dawes, W. R. (2010), Impacts of climate change on groundwater in Australia: a sensitivity analysis of recharge, Hydrogeology Journal, Vol. 18, pp. 1625–1638.

Mollema, Pauline N. and Antonellini, Marco (2013), Seasonal variation in natural recharge of coastal aquifers, Hydrogeology Journal, Volume 21, Issue 4, pp. 787-797.

National Environment Policy (2004), Ministry of Environment and Forests, Government of India, p. 38.

Neukum, Christoph and Azzam, Rafig (2012), Impact of climate change on groundwater recharge in a small catchment in the Black Forest, Germany, Hydrogeology Journal, Volume 20, Issue 3, pp. 547-560.

Okkonen, Jarkko, Jyrkama, Mikko and Kløve, Bjørn (2010), A conceptual approach for assessing the impact of climate change on groundwater and related surface waters in cold regions (Finland), Hydrogeology Journal, Vol. 18, pp. 429–439.

Oude Essink, G. H. P., van Baaren, E. S. and de Louw, P. G. B. (2010), Effects of climate change on coastal groundwater systems: A modeling study in the Netherlands, Water Resources Research, Vol. 46.

Payne, Dorothy F. (2010), Effects of climate change on saltwater intrusion at Hilton Head Island, SC. U.S.A., SWIM21 - 21st Salt Water Intrusion Meeting, 21-26 June 2010, Azores, Portugal, pp. 293-296.

Ranjan, P., Kazamaa, S. and Sawamoto, M. (2006), Effects of climate change on coastal fresh groundwater resources, Global Environmental Change, Vol. 16, pp. 388–399.

Raposo, Juan Ramón, Dafonte, Jorge and Molinero, Jorge (2013), Assessing the impact of future climate change on groundwater recharge in Galicia-Costa, Spain, Hydrogeology Journal, Volume 21, Issue 2, pp. 459-479.

Ravindranath, N. H., Joshi, N. V., Sukumar, R. and Saxena, A. (2005), Impact of climate change on forests in India, Current Science, Vol. 90, pp. 354–361.

Richardson, C. W. and Wright, D. A. (1984), WGEN: A model for generating daily weather variables. ARS-8. Agricultural Research Service, US Department of Agriculture.

Rosenberg, N. J., Epstein, D. J., Wang, D., Vail, L., Srinivasan, R. and Arnold, J. G. (1999), Possible impacts of global warming on the hydrology of the Ogallala Aquifer Region, Climatic Change Vol. 42, pp. 677–692.

Rozell, Daniel J. and Wong, Teng-fong (2010), Effects of climate change on groundwater resources at Shelter Island, New York State, USA, Hydrogeology Journal, Vol. 18, pp. 1657-1665.

Rupa Kumar, K. (2005), High-resolution climate change scenarios for India for the 21st century, Current Science, Vol. 90, pp. 334–345.

Sathaye, Jayant, Shukla, P. R. and Ravindranath, N. H. (2006), Climate change, sustainable development and India: Global and national concerns, Current Science, Vol. 90, Number 3, 10 February 2006, pp. 314-325.

Scibek, J. and Allen, D. M. (2006), Modeled impacts of predicted climate change on recharge and groundwater levels, Water Resources Research, Vol. 42, W11405, 18p.

Shah, Tushaar (2009), Climate change and groundwater: India's opportunities for mitigation and adaptation, Environmental Research Letter 4, IOP Publishing Ltd, UK.

Sherif, Mohsen M. and Singh, Vijay P. (1999), Effect of climate change on sea water intrusion in coastal aquifers, Hydrological Processes, Vol. 13, pp. 1277-1287.

Singh, R. D. and C. P. Kumar (2010), Impact of Climate Change on Groundwater Resources, Proceedings of 2nd National Ground Water Congress, 22nd March 2010, New Delhi, pp. 332-350.

Taylor B. (1997), Climate change scenarios for British Columbia and Yukon, In: Taylor E, Taylor B. (eds), Responding to global climate change in British Columbia and Yukon, Volume I of the Canada country study: climate impacts and adaptation, Environment Canada and BC Ministry of Environment, Lands and Parks.

Toews, Michael W. (2007), Modelling Climate Change Impacts on Groundwater Recharge an a Semi-Arid Region, Southern Okanagan, British Columbia, A thesis submitted in partial fulfillment of the requirements for the degree of Master of Science in the Department of Earth Sciences, Simon Fraser University.

UNFCCC (2004), India's Initial National Communications to the United Nations Framework Convention on Climate Change, Ministry of Environment and Forests, New Delhi.

Vandenbohede, Alexander and Lebbe, Luc (2010), Impact of climate change on the phreatic aquifer of the western Belgian coastal plain, SWIM21 - 21st Salt Water Intrusion Meeting, 21-26 June 2010, Azores, Portugal, pp. 297-300.

Vaccaro, J. J. (1992), Sensitivity of groundwater recharge estimates to climate variability and change, Columbia Plateau, Washington, Journal of Geophysical Research, Vol. 97 (D3), pp. 2821-2833.

Woldeamlak, S. T., Batelaan, O. and De Smedt, F. (2007), Effects of climate change on the groundwater system in the Grote-Nete catchment, Belgium, Hydrogeology Journal, Vol. 15, Number 5, August 2007, pp. 891-901.

Yihdego, Yohannes and Webb, John A. (2011), Modeling of bore hydrographs to determine the impact of climate and land-use change in a temperate subhumid region of southeastern Australia, Hydrogeology Journal, Volume 19, Issue 4, pp. 877-887.

Zhou, Yu, Zwahlen, François, Wang, Yanxin and Li, Yilian (2010), Impact of climate change on irrigation requirements in terms of groundwater resources, Hydrogeology Journal, Vol. 18, pp. 1571–1582.